YOUR KNOWLEDGE HAS VALUE

Mohammed Obeid

Characteristic of Wollastonite Synthesized from Local Raw Materials

GRIN Publishing

Bibliographic information published by the German National Library:

The German National Library lists this publication in the National Bibliography;
detailed bibliographic data are available on the Internet at http://dnb.dnb.de .

Imprint:

Copyright © 2014 GRIN Verlag GmbH
Print and binding: Books on Demand GmbH, Norderstedt Germany
ISBN: 978-3-656-93249-9

This book at GRIN:

http://www.grin.com/en/e-book/294570/characteristic-of-wollastonite-synthesized-
from-local-raw-materials

GRIN - Your knowledge has value

Since its foundation in 1998, GRIN has specialized in publishing academic texts by students, college teachers and other academics as e-book and printed book. The website www.grin.com is an ideal platform for presenting term papers, final papers, scientific essays, dissertations and specialist books.

Visit us on the internet:

http://www.grin.com/

http://www.facebook.com/grincom

http://www.twitter.com/grin_com

EXPLORE
CENTRE OF PROFESSIONAL RESEARCH PUBLICATIONS
International
Journal of Engineering & Technology
ISSN 2049-3444

International Journal of Engineering and Technology Volume 4 No. 7, July, 2014

Characteristic of Wollastonite Synthesized from Local Raw Materials

Majid Muhi Shukur, Elham Abd Al-Majeed , Mohammed Maitham Obied

College of Materials Engineering, University of Babylon 51002, Babylon, Iraq

ABSTRACT

In this study, wollastonite ($CaSiO_3$) has been synthesized by solid state reaction method at temperature range of 1050-1300 °C from local raw materials e.g. silica sand and limestone with addition of small amount of B_2O_3 as a mineralizer to activate the reaction process during sintering. The resulting products are investigated employing XRD and SEM techniques. β-wollastonite was obtained at 1050 °C and transformed to pseudowollastonite (α-$CaSiO_3$) at 1150 °C. Physical and thermal properties have been evaluated. The batch sintered at 1250 °C showed the highest value of density of about 1.98 g/cm³ with high shrinkage rate as compared with other batches. The results also revealed that thermal expansion coefficient is more compatible with results obtained from natural wollastonite.

Keywords: *wollastonite, solid state reaction method, XRD, SEM, B_2O_3 addition*

1. INTRODUCTION

Wollastonite is natural occurring calcium metasilicate with the chemical formula CaSiO3 [1]. It has a combination of properties, such as low dielectric constant, low dielectric loss, thermal stability, low thermal expansion and low thermal conductivity, hence is used in ceramic fabrication, medical material for artificial bone and dental roots, high frequency insulators, filler material in resins and plastics, civil construction, metallurgy, paints and frictional products [2,3,4,5]. This situation has tempted a number of researchers to produce it synthetically [6].

Wollastonite crystallizes in three polymorphic forms, low temperature triclinic [1T] and monoclinic or so called parawollastonite [2M] and high temperature form pseudowollastonite which occur in a pseudo-hexagonal form. The conversion of the low temperature to high temperature form takes place at 1125 °C [7]. Wollastonite can be manufactured by melting suitable raw materials such as, CaO, Ca(OH)2, CaCO3, CaSO4 and SiO2 as quartz sand and slag rising from the phosphorous plants, followed by controlled crystallization of the glasses or by sol-gel method. The main drawbacks of the melting method are the multi-step preparation procedure and the relatively high processing temperatures (about1500 °C).The sol-gel method offers lower processing temperatures (they do not exceed 1200°C), however it is expensive and has low efficiency. While in solid state method, synthetic wollastonite is crystallized from a homogenous mixture of inexpensive raw materials in a 1:1 molecular ratio when heated to a temperature below its melting point with addition of a fluxing agent [8, 9].

In this work, solid state reaction method has been employed for synthesizing of wollastonite from local raw materials e.g. silica sand and limestone as a source for CaO where they distributed abundantly in many regions of Iraq.

2. EXPERIMENTAL PROCEDURE

2.1 Preparation of Specimens

Wollastonite has been synthesized by solid state method based on a mixture of silica sand (99.56 wt%) and CaO extracted from local limestone after its calcination at 950°C for 5 hours with the addition of 4 wt% of H3BO3 to provide 2.24 wt% of B2O3. Mixtures of a molar ratio of CaO:SiO2 equal to one were first milled using planetary ball mill (SFM-1, QM-3SP2) which runs at 300 rpm for 15 hours using distilled water as a dispersive media with an addition of 2 wt% PVA as a binder. The mixture was oven dried at 80 °C for 24 h and calcined at 710 °C for 3 h in air atmosphere. Subsequently, the powder mixture was compacted at 150 MPa [10] and sintered at different temperatures of 1050, 1150, 1250 and 1300 °C for 2 h with an average heating rate of 7 °C/min.

2.2 Characterizations

After milling and drying, Bettersize2000 was employed to determine the milled powder of raw materials in which particle size measurements were made at a relatively low concentration of 0.5 g powder dispersed in 750 ml of distilled water as shown in Fig.1 below. All samples obtained from the fired batches were finely grounded and scanned using x-ray diffractometer (Shimadzo, 6000) at room temperature using Cukα radiation (λ = 1.5405 Å), and a scanning speed of 5°/min from 20° to 50° of 2Θ (Bragg angle) and 40 kV/30 mA as an applied power to estimate the crystallized phases.

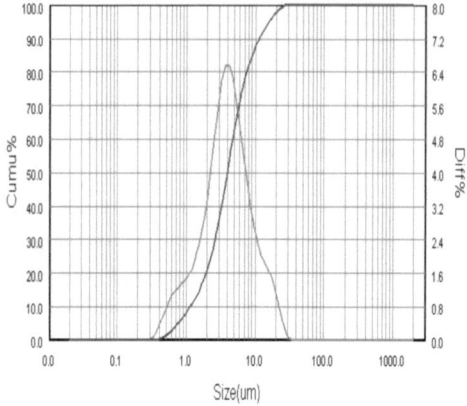

Fig.1 Particle size analysis of the mixed powder

The microstructures were observed using the scanning electron microscopy (SEM) (VEGAII/TESCAN XMV). All samples were finely polished and then coated with a thin layer of gold by sputtering deposition technique (EMITECH, K450X, UK), prior to being scanned using SEM instrument to produce images with a magnification of (X2000).

The bulk density was determined according to ASTM C373 [11] and the percentage linear firing shrinkage was calculated according to the formula:

$$Shrinkage\ \% = [(L_o - L)/\ L_o\ ]\ x\ 100$$

Where, L_o and L are the lengths of the specimen before and after firing process respectively [12].

Thermal expansion coefficient of the fired samples were measured at a temperature range of 25-600 °C using manual dilatometer (Quickline – 05) according to the expression below:

$$\alpha = 1/L_o\ .\ \Delta L/\Delta T$$

Where, L_o is the length of the test piece at ambient temperature, ΔL is the increase in the length per unit change in temperature and ΔT is the difference in temperature [13].

3. RESULTS AND DISCUSSIONS

3.1 X-ray diffraction (XRD)

Fig.2 shows the XRD patterns of the crystallization behavior of powder sintered at different temperatures. Samples sintered at 1050 °C comprised of monoclinic (β-CaSiO$_3$) at a

diffraction angle of 29.898° (2.98 A°), with little amount of unreacted quartz at 2Θ equal to 26.566°. Meanwhile, some portion of SiO$_2$ had been converted into its high temperature polymorph tridymite at 2Θ of 21.776°. On further increasing in firing temperature to 1150 °C, pseudowollastonite (α-CaSiO$_3$) could be observed at a diffraction angle of 27.479° (3.24 A°) and continued its crystallization to give the highest value at 1250 °C associated with cristobalite. At 1300 °C, the pattern is similar to that of 1250 °C but calcite is appeared at 2Θ equal to 29.369° (3.03 A°) which may be related to the lack of dwelling time at 1300 °C.

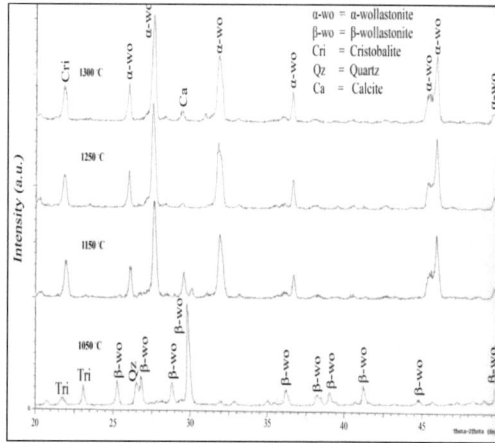

Fig.2 XRD patterns of batches sintered at different temperatures

3.2 SEM Micrographs

The microstructures of CaSiO$_3$ sintered batches are demonstrated in Fig.3 below. Picture (a) represents the morphological analysis of the specimen sintered at 1050 °C, it can be seen that the specimen has a porous structure with granular particle shape. On contrary, the grains as shown in picture (b) were larger with subspherical of interconnected pores and smooth surface which seems more consolidate. In picture (c) the microstructure shows an appropriate grain growth of a large size which increased as a function of temperature. While at 1300 °C, the microstructure appears quite similar in morphology to the batch sintered at 1250 °C. This may be related to the lack of dwelling time at 1300 °C.

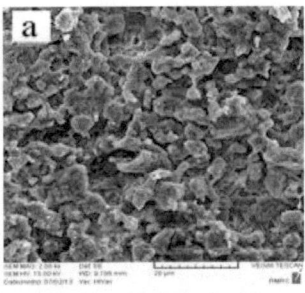

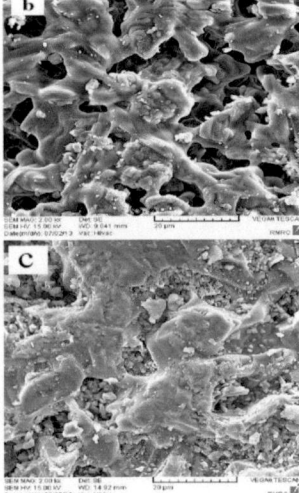

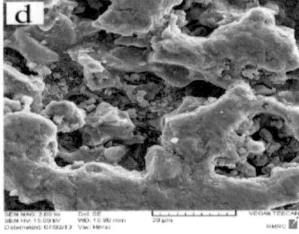

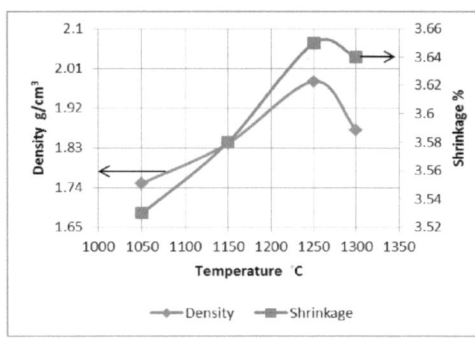

Fig.3 SEM micrographs of batch sintered at (a)1050,(b) 1150, (c) 1250 and (d) 1300 ˚C

3.3 Properties Studies

The density, shrinkage and thermal expansion coefficient have been studied for samples sintered at different temperatures e.g. 1050, 1150, 1250 and 1300 ˚C as shown in table.1. It indicates that, with increasing in sintering temperature the density was increased. It is also noted that samples contained 2.24 wt% of B_2O_3 have density higher than that contained no addition where B_2O_3 form a liquid phase in early stage of sintering temperature and promote grain growth [14]. As the density of the samples increased the rate of shrinkage is also increased as shown in Fig.4.

Specimens sintered at 1050 ˚C for 2 h, where β-CaSiO$_3$ is appeared as a major phase which exhibit the lowest value of the coefficient of thermal expansion of about 1.08×10^{-6}/˚C at 100 ˚C and increased linearly to reach it highest value of about 6.96×10^{-6}/˚C at 600 ˚C as shown in Fig.5. So far, pseudowollastonite (α-CaSiO$_3$) is appeared at 1150 ˚C and continued in its crystallization up to 1300 ˚C. As a matter of fact, the lines related to batches sintered at 1150, 1250 and 1300 ˚C are closely compatible where the values of coefficient of thermal expansion are in the range of 11.36-11.67×10^{-6}/˚C at 600 ˚C.

Table.1: The Properties of the Synthetic Wollastonite

Temp.(˚C)	*B (g/cm^3)	Shrinkage (%)	*CTE (x 10^{-6}/˚C) at 600 ˚C
1050	1.75	3.53	6.96
1150	1.84	3.58	11.44
1250	1.98	3.65	11.31
1300	1.87	3.64	11.67

*B = Bulk density

*CTE = Coefficient of thermal expansion

Fig.4 effect of temperature on density and shrinkage

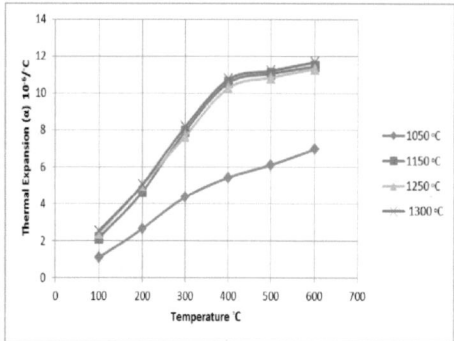

Fig.5 Thermal expansion of the sintered batches

4. CONCLUSIONS

Synthetic wollastonite (CaSiO3) has been synthesized from local raw materials by solid state method at different temperatures. Low temperature wollastonite (β- CaSiO3) was achieved at 1050 °C with a density of 1.75 g/cm3. As the temperature increased to 1150 °C, pseudowollastonite can be observed and continued its crystallization up to 1250 °C with a maximum density of about 1.98 g/cm3. Linear thermal expansion for samples sintered at 1050 °C is more compatible with natural wollastonite.

Acknowledgements

Words of appreciation are presented to the staff of the laboratory of ceramic and composite materials, laboratory of electrochemical engineering and Al-Noora factory for their cooperation and invaluable experience.

REFERENCES

[1]. M. S. Nizami, "Characterization of a mineral synthesized by availing silica from plant source," *J. mater. Sci. Technol.*, vol. 19, no. 6, pp. 599-603, 2003.

[2]. R. P. Sreekanth Chakradhar, B. M. Nagabhushna, G. T. Chandrappa, K. P. Ramesh and J. L. Rao, "Solution Combustion Derived Nanocrystalline Macroporous Wollastonite Ceramics," *Mater. Chemistry and Physics*, vol. 95, no. 1, pp. 169-175, 2006.

[3]. H. Ming, Z. S. Ren, Z. X. Hua, "Characterization and analysis of $CaO-SiO_2-B_2O_3$ ternary system ceramics," *J. Mater. Sci: Mater Electron*, vol. 22, no. 4, pp. 389-393, 2011.

[4]. S. Vichaphund, M. Kitiwan, D. Atong, P. Thavorniti, "Microwave synthesis of wollastonite powder from eggshells" J. of the European Ceramic Society, vol. 31, pp. 2435-2440, 2011.

[5]. K. Lin, J. Chang, J. Lu, "Synthesis of wollastonite nanowires via hydrothermal microemulsion methods" Materials Letters, vol. 60, pp. 3007-3010, 2006.

[6]. S. H. Javed, S. Naveed, S. Tajwar, M. Shafaq, M. Kazmi and N. Feroze, "SYNTHESIS OF WOLLASTONITE FROM AGRO-WASTE," *J. Agric. Res.*, vol. 47, no. 3, pp. 319- 327, 2009.

[7]. A. Yazdani, H. R. Rezaie and H. Ghassai, "Investigation of hydrothermal synthesis of wollastonite using silica and nano silica at different pressures," *J. of Ceramic Processing Research*, vol. 11, no. 3, pp. 348-353, 2010.

[8]. J. Podporska, M. Balzewicz, B. Trybulska and L. Zych, "A novel ceramic material with medical application," *Processing and Application of Ceramics*, vol. 2, no. 1, pp. 19-22, 2008.

[9]. R. H. MacDonald, "Various industrial mineral commodities in Nova Scotia," *Mines and Energy Branches, Economic Geology Series 92-1*, 1992.

[10]. A. Harabi and S. Chehlatt, "Preparation of highly resistant wollastonite bioceramics using local raw materials," *J. Therm. Anal. Calorim.*, vol. 111, no. 1, pp. 203-211, 2013.

[11]. ASTM C373-88, "Standard Test Method for Water Absorption, Bulk Density, Apparent Porosity, and Apparent Specific Gravity of Fired Whiteware Products," *Annual Book of Standards*, ASTM, West Conshohocken, PA, 1999.

[12]. B. Ertuğ,"SINTERINGAPPLICATIONS," *InTech*, Rijeka, Croatia, 2013.

[13]. M. Kutz, "Handbook of Materials Selection," *John Wiley & Sons*, New York, 2002.

[14]. H. -P. Wang, S. -Q. Xu, S. -Q. Lu, S. -L. Zhao and B. -L. Wang, "Dielectric properties and microstructures of $CaSiO_3$ ceramics with B_2O_3 addition," *Ceramics International*, vol. 35, no. 7, pp. 2715-2718, 2009.